P9-CAY-816

3 1611 00091 7481

GOVERNORS STATE UNIVERSITY
UNIVERSITY PARK
IL 60466

OKLAHOMA STATE UNIVERSITY
UNIVERSITY PARK

SOUTH AMERICAN ANIMALS

CAROLINE ARNOLD

Morrow Junior Books • New York

PHOTO CREDITS: Arthur Arnold: pages 28–29, 36–37, 46–47; Caroline Arnold: pages 38–39, 42–43; Matthew Arnold: pages 1, 14–15, 30–31, 34–35; Heinz Plenge: jacket and pages 2–3, 8–9, 10–11, 16–17, 18–19, 20–21, 22–23, 48; Dr. R. Ridgely/VIREO: pages 32–33; Adina Sales: pages 12–13, 24–25; Thomas Struhsaker: pages 4–5, 40–41; Michael Wallace: pages 26–27; Doug Wechsler: pages 6–7, 44–45.

The text type is 17-point Weiss.   Book design by Christy Hale

Copyright © 1999 by Caroline Arnold

All rights reserved. No part of this book may be reproduced or utilized in any form or by any means, electronic or mechanical, including photocopying, recording, or by any information storage and retrieval system, without permission in writing from the Publisher.   Published by Morrow Junior Books, a division of William Morrow and Company, Inc., 1350 Avenue of the Americas, New York, NY 10019   www.williammorrow.com

Printed in Hong Kong by South China Printing Company (1988) Ltd.   10 9 8 7 6 5 4 3 2 1

Library of Congress Cataloging-in-Publication Data
Arnold, Caroline.
South American animals/Caroline Arnold.
p. cm.
Summary: Discusses the variety of animals found in the rain forests, mountains, grasslands, and coastal regions of South America, including the birds, mammals, reptiles, and amphibians.
ISBN 0-688-15564-2 (trade)—ISBN 0-688-15565-0 (library)
1. Zoology—South America—Juvenile literature. [1. Zoology—South America.] I. Title. QL235.A75 1999
591.98—dc21   98-7669   CIP AC

MAT-CTR. QL 235 .A75 1999

Arnold, Caroline.

South American animals

CONTENTS

A BIG CONTINENT

IN THE DAPPLED SHADE OF A SOUTH AMERICAN rain forest a young ocelot creeps across the ground. In the branches above birds sing and insects buzz. At a nearby river turtles rest on a fallen log.

South America is the world's fourth-largest continent and is home to an enormous number of wild animals. They live in forests, mountains, and grasslands, and at the water's edge. Each place, or habitat, provides food and shelter for many kinds of animals. Let's discover how some South American animals live and find the things they need to survive.

FORESTS

Rain forests
Other forests

THE NORTHERN HALF OF SOUTH America is in the warm, tropical region of the world. Much of it is covered by the huge Amazon rain forest, where moisture and warm temperatures help plants and trees grow. From the tallest branches at the tops of the trees to the shaded forest floor, each layer of the rain forest provides food and shelter for a wide variety of animals.

You can also see wildlife in other, less rainy South American forests.

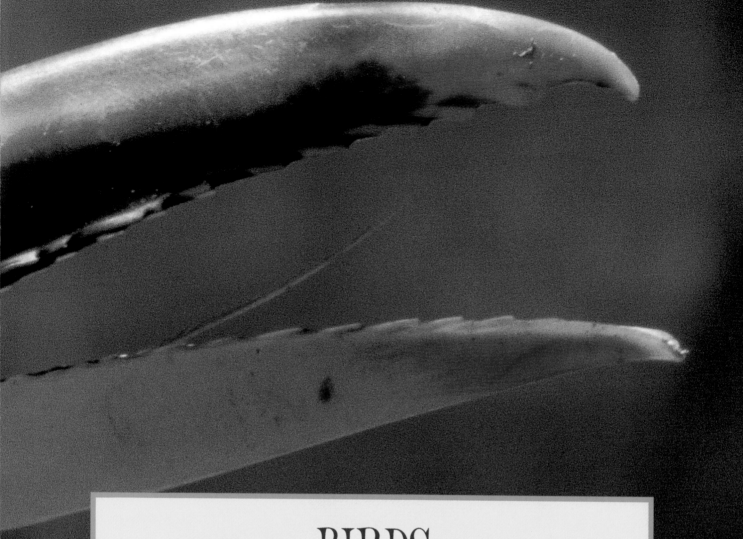

BIRDS

NO OTHER CONTINENT HAS AS MANY DIFFERENT kinds, or species, of birds as South America. Noisy parrots, tiny hummingbirds, and colorful toucans are some of the birds you can see in the rain forest.

Birds' beaks help them to eat their favorite foods. Parrots use their strong beaks to crack nuts, slice open fruit—and hold on to branches as they climb! Hummingbirds hover in the air and sip nectar from flowers with their long, narrow beaks. And the toucans' long bills help them to reach fruits and berries.

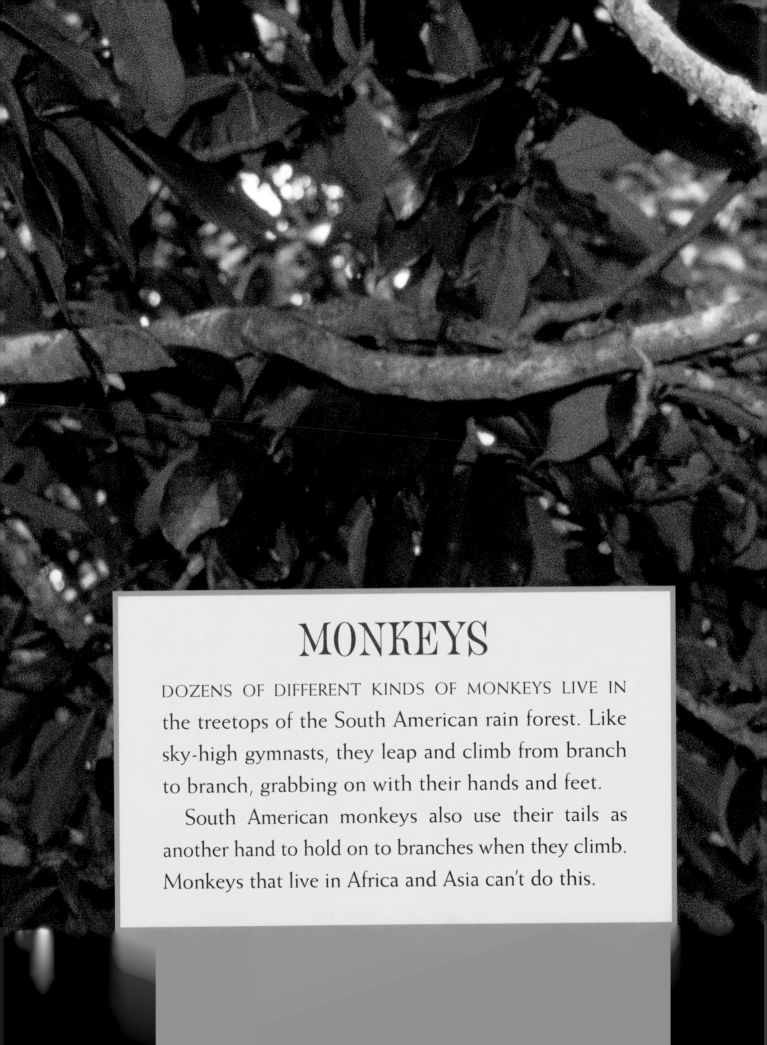

MONKEYS

DOZENS OF DIFFERENT KINDS OF MONKEYS LIVE IN the treetops of the South American rain forest. Like sky-high gymnasts, they leap and climb from branch to branch, grabbing on with their hands and feet.

South American monkeys also use their tails as another hand to hold on to branches when they climb. Monkeys that live in Africa and Asia can't do this.

12

SLOTHS

SLOTHS ARE SLOW-MOVING ANIMALS THAT ARE found only in Central and South America. Their actions are so slow that sloths seem to be part of a slow-motion movie. This may make it harder for hunters such as jaguars to notice them.

Sloths spend most of their lives in trees—and are experts at hanging upside down. As a sloth climbs among branches looking for leaves to eat, long, sharp nails at the ends of its toes help it hold on. A sloth can also turn its head almost completely around. Would you like to be able to swivel your head so that you could see behind you?

SNAKES

MANY KINDS OF SNAKES LIVE IN THE SOUTH American rain forest. One of them is the anaconda, the biggest snake in the world. Some anacondas are more than thirty-seven feet long. That's the length of a school bus!

All snakes are hunters. They eat birds, insects, and small animals. Each snake has its own way of finding food. Some slither along the forest floor. Others swim after their prey. And some, like the emerald tree boa, climb trees. Its thin, scaly body is good for gripping branches as it searches for birds to eat. When an emerald tree boa rests, it just wraps itself around a branch and goes to sleep.

FROGS

HUNDREDS OF DIFFERENT KINDS OF FROGS LIVE IN THE rain forest. Some of them are so colorful that they look as if they were dipped in paint. All of the many species of poison dart frogs have bright markings. This warns other animals, *Don't eat me, because I will make you sick.*

Frogs have moist skin and need to keep it wet. This makes the damp rain forest an ideal home for them. These long-legged amphibians are at home in and out of the water. Their powerful back legs are good for both jumping and swimming. Can you kick your legs like a frog when you jump or swim?

JAGUARS

WHEN A JAGUAR WALKS THE FOREST FLOOR, OTHER animals get out of the way. This meat eater is the largest predator in South America. It can grow up to eight feet long. Jaguars hunt at night and can see well in the dark. They feed on alligators, tapirs, and other large forest animals.

Pumas and ocelots are some of the many other South American wild cats. Whether big or small, all cats are excellent hunters. Good eyesight and hearing, quick reflexes, and sharp teeth and claws help them catch the animals that are their prey.

ARMADILLOS

THE ARMADILLO GETS ITS NAME FROM ITS SCALY SKIN, which makes this unusual mammal look as if it is wearing armor. The hard covering protects an armadillo from most predators.

Armadillos live only in Central and South America and in the southern United States. There are twenty species of armadillos. All of them are insect eaters. Armadillos find their food by smelling for it with their long noses. Then they use their sharp-clawed front feet to dig it up.

TAPIRS

THE LARGEST ANIMALS FOUND IN THE RAIN FOREST ARE tapirs. These stout, short-legged mammals may grow to be more than six feet long and weigh up to seven hundred pounds. Their skin is covered with short, stiff hairs.

Tapirs have a good sense of smell and use their long, flexible snouts to find leaves, buds, and fruits to eat. They usually feed at night and rest during the day.

Tapirs are related to rhinoceroses and horses. Like rhinos, they spend much of their time in water.

TARANTULAS

TARANTULAS ARE THE BIGGEST SPIDERS IN THE WORLD. One species that lives in South America is larger than your hand. Its body is more than three inches long, and each of its eight legs measures five inches.

A tarantula's home is a burrow in the ground. The tarantula sleeps there during the day and goes hunting at night. Its dark color makes it hard to see among the leaves on the forest floor. Walking on the tips of its long, hairy legs, the tarantula looks for insects. When it finds an insect, the tarantula kills it with a poisonous bite. A tarantula bite cannot kill a person, but it can be painful. Unless tarantulas feel threatened, though, they usually do not harm people.

MOUNTAINS

THE ANDES MOUNTAINS FORM an unbroken ridge along the western side of South America. Many of the jagged peaks are more than three miles high. Even near the equator they are always covered with snow. Llamas and guanacos live on the high, dry plains that lie between these mountain ridges. Above them condors soar on ten-foot-wide wings. These huge vultures are the largest flying birds in the world. Bears, birds, and other animals live on lower mountain slopes.

The Andes are so tall that they stop rain clouds from reaching the other side. Between the Andes Mountains and the Pacific Ocean lies the driest desert in the world. Parts of southern South America that get a little more rain are semidesert regions.

Mountains
Deserts
Semideserts

GUANACOS

IN HIGH MOUNTAIN MEADOWS GUANACOS FEAST ON fresh green grass. Like their relative the camel, guanacos can go for many days without drinking. They get most of the water they need from the food they eat. Guanacos are like camels in another way, too—when they fight each other, they spit.

Other members of the camel family that live in South America are vicuñas, which are wild, and llamas and alpacas, which are domestic animals. People use llamas to carry things. The warm wool of llamas and alpacas is good for making blankets and clothing.

SPECTACLED BEARS

WITH DARK FUR AND WHITE LINES ON ITS FACE, THE spectacled bear almost looks as if it is wearing glasses. This bear lives in the forests and foothills of the Andes Mountains. Its sharp claws help it get a good grip when climbing tall trees in search of leaves and fruit. It also eats insects and occasionally the meat of small animals.

Spectacled bears are the only bears in South America. They are smaller than most other kinds of bears. When an adult spectacled bear stands on its back legs, it is about as tall as a grown person.

GRASSLANDS

VAST GRASSLANDS EXTEND ACROSS much of southern South America. There are also grasslands in the central and northern parts of the continent. Ranchers use these broad plains for raising cattle. But many wild animals live there, too, including large birds called rheas. These relatives of the ostrich cannot fly—they run instead.

Grasslands

34 PUDU

DEER

SEVERAL SPECIES OF DEER LIVE IN SOUTH AMERICA. ALL of them are small compared to their North American relatives.

The pudu, which makes its home in forest regions of the south, is the smallest deer in the world. It is only a foot tall! Pudus have short legs and tiny tails. These hoofed animals live in small herds and are active mainly at dawn and dusk.

All deer are plant eaters and have strong, flat teeth that are good for chewing. Which of your teeth do you use when you chew?

FOXES

THE ARGENTINE GRAY FOX, THE COLPEO FOX, AND the crab-eating fox are among several kinds of foxes that live in South America. Sharp eyes, excellent hearing, and quick reflexes make all foxes expert hunters. Foxes will eat almost anything they can catch, including mice, birds, fish, and worms. A fox usually pounces on a victim and then grabs it with sharp teeth.

As a fox runs, its long, bushy tail helps it to keep its balance.

MARAS

MARAS ARE AMONG A GREAT VARIETY OF SOUTH American rodents. Like all rodents, they have strong front teeth. These are good for cutting and gnawing the plants that are their food. Although a mara looks like a large rabbit, it is really a cavy, a group of animals that includes guinea pigs.

Maras live on the dry plains of southern South America. They have an unusual way of getting around—as they leap across the grasslands, they bounce on all four legs at once. In this way maras can jump more than six feet high. How high can you jump?

AT THE WATER'S EDGE

At the water's edge

FROM THE TROPICAL NORTH COASTS to the chilly southern seas, South America's shorelines are abundant with wildlife. Seabirds,

turtles, seals, and sea lions are some of the animals that make their home at the edge of the sea.

Animals also live near South American rivers and lakes. In tropical regions caimans sometimes crawl out of the water onto sandy riverbanks to rest. These relatives of alligators feed on fish and other small animals.

FLAMINGOS

WITH LEGS LIKE STILTS AND LONG PINK NECKS, flamingos are among the tallest and most beautiful of all birds. Several kinds of flamingos live in South America. Some live along the edge of shallow ocean bays in the tropics. Others are found in lakes high in the Andes Mountains.

A flamingo finds food by straining water through its large beak. The beak's comblike edges catch pieces of food in the same way you might use a tea strainer to catch tea leaves.

A flamingo sleeps standing up. It folds one leg under its body, tucks its head under its wing, and goes to sleep. Can you close your eyes and stand on one leg?

SEA TURTLES

IN THE WARM, SHALLOW WATER OFF THE EAST COAST of South America you can see huge green sea turtles. These large reptiles can be three feet across and weigh more than two hundred pounds.

At nesting time a female sea turtle returns to the beach where she was born. She can recognize it by its smell. There she digs a hole in the sand and lays a pile of round, white eggs. Then she goes back to the ocean. When the baby turtles hatch, they too will head for the ocean.

Sea turtles are expert swimmers. Their broad, flat feet are like flippers. The turtles use them to push themselves through the water. Do you sometimes wear flippers to help you swim fast?

GREEN SEA TURTLE   45

PENGUINS

MOST PEOPLE THINK PENGUINS ARE COLD-WEATHER BIRDS. Some kinds of penguins *do* live in ice and snow. But many others live in warmer climates. Each summer millions of penguins come to South American shores to build nests, lay eggs, and raise their babies. Then at summer's end all the penguins return to the sea until the next nesting season.

Penguins are built for a life at sea. These birds cannot fly, but they are excellent swimmers. They use their wings like paddles and steer with their feet. Tightly packed oily feathers help keep their bodies dry and warm.

DISAPPEARING WILDLIFE

THE AMAZON RAIN FOREST IS HOME TO A greater variety of animals than any other place on earth. However, as the rain forest and other wild lands in South America disappear, the number of animals that live in those places grows smaller. By learning how South American animals live at home in the wild, we can find out ways to help them and give them a better chance to survive in the future.